Published by Smart Apple Media
1980 Lookout Drive, North Mankato, MN 56003

Designed by Stephanie Blumenthal
Production Design by Kathy Petelinsek

Photographs by Maslowski Wildlife Photography

Copyright © 2002 Smart Apple Media.
International copyrights reserved in all countries.
No part of this book may be reproduced in any form without
written permission from the publisher.

Library of Congress Cataloging-in-Publication Data

Maslowski, Stephen. Birds in spring / by Steve Maslowski with Adele Richardson.
p. cm. — (Through the seasons ; 1)
Summary: Focuses on the behaviors of various kinds of birds as they return to northern
areas in the spring, claim territory, court, build nests, and lay eggs.
ISBN 1-58340-056-7
1. Birds—Nests—Juvenile literature. 2. Birds—Behavior—Juvenile literature.
[1. Birds—Nests. 2. Birds—Behavior.] I. Richardson, Adele. II. Title.

QL675 .M46   2001
598.156—dc21   99-046949

First Edition

2 4 6 8 9 7 5 3 1

THROUGH THE SEASONS

BIRDS IN SPRING

Text by Steve Maslowski with Adele Richardson
Photographs by Maslowski Wildlife Photography

SMART APPLE MEDIA

As spring approaches, warmer weather begins to melt the snow that covers much of North America. The ground becomes soft and muddy, and trees and plants everywhere sprout new leaves. Each day, the sun shines a little longer. The season of spring begins on the 20th day of March. This is a very busy time of year for birds.

Day and night are exactly the same length—12 hours—on the first day of spring.

Many birds spend the winter in southern regions, away from the cold and snow. As the seasons change, these birds fly back to their homes in the North. Once they complete the trip, they spend nearly every hour of the day guarding territories and finding mates. Soon, nests of all shapes and sizes are built in trees and bushes across the continent.

Vireo collecting nesting material

Male birds are the first to arrive in the North. They often show up days or even weeks before the females. Usually, birds return to the same areas in which they were hatched. Geese are among the first birds to return to their northern homes in the spring. Sometimes they show up so early in the season that the snow and ice have not yet melted.

The United States' record low spring temperature was set on April 5, 1945, when it dropped to –36° F (–38° C) in New Mexico.

Flycatcher nest (top); young bluebirds (bottom)

Robins and killdeers also fly north early. Many times, these birds will see bright sunshine one day and cold snow the next. This type of weather often makes food easier to find. Melting snow soaks into the ground, making it soft. Many birds eat the earthworms that are forced to the surface by the water.

The male woodcock performs tricks to attract a mate. In the evening, he sings and flies high into the air. He then suddenly closes his wings and falls to the ground, landing close to the spot where the female is sitting. Male woodcocks have a special song that they sing only while courting females.

Killdeer

Once the male birds reach their spring homes, they claim a territory. Birds consider a territory their own land and air—a place where food can always be found. Territories should be in safe places. This is important because a bird's territory is where the bird and its partner will build a nest to raise a family, or brood, of baby birds.

The size of a bird's territory depends on the type of bird. Predators hunt other animals for food, so they need more space. Hawks, eagles, and falcons are all predators that need large hunting grounds. The territory of just one of these birds may cover several miles.

Chickadee chicks waiting to be fed

Smaller birds, such as robins and blue jays, don't need as much land. Because they feed mainly on worms and insects, their territory may cover only a few acres. Sometimes it's as small as a person's backyard.

When the female birds arrive, the air is filled with the sound of males'

The oriole builds a unique nest. The nest hangs down from the branch of a tree and swings when the wind blows. To build this pouch-like nest, the oriole uses hair, thread, and thin grass. The nest is very deep with a small opening at the top.

birdsongs. Male birds do many things to catch the attention of females. They may sing, dance around, or show off their colorful feathers. These displays are all part of the courting process. Male birds sing not only to win females, but also to warn other males to stay out of their territory. If one male flies into another's territory, he will be pecked at and chased away.

The killdeer usually lays four or five gray eggs with dark spots on them. Once they have been laid in the nest, the eggs must be kept warm. The male and female killdeers take turns doing this. One will sit on the eggs while the other searches for food. Killdeer eggs take about 12 to 16 days to hatch.

Male bobolink singing

But not all male birds sing to attract females. The ruffed grouse, for example, makes a loud drumming sound instead by beating his wings while sitting on a log. Male woodpeckers also make a drumming noise. They find a hollow tree or a telephone pole, then peck away with their beaks.

The bald eagle builds its nest high in a treetop or on a mountain. An eagle nest can be gigantic—some have been measured at up to eight feet (2.5 m) across and 12 feet (3.6 m) deep. Bald eagle eggs take between 31 and 46 days to hatch.

Ruffed grouse drumming

Bluebird nest with eggs

After male and female birds are paired, they begin building a nest. The females do most of this work, starting with picking out the exact spot on which to build the nest.

Nest building is not something that birds have to learn; they are hatched with this ability. Birds can build a perfect nest even if they have never seen one before.

Many birds build a small, bowl-shaped nest. The robin is one of these birds. The female robin picks out a spot in a tree or bush, or perhaps even on a window sill. She then flies back and forth to the nest site, carrying twigs and grass. These materials will form the bottom of the nest.

To hold the nest together, the robin uses mud. If she cannot find mud, she will pick up dirt with her beak and dip it in some water. As the mud dries, it holds the twigs and grass together. Then the robin can

Robin parent and chicks

build up the sides. She uses more twigs and grass, or string and pieces of cloth if she can find any. When the nest is finished, the robin sits inside and turns around to smooth the nest walls. She then adds some soft grass to the bottom as padding for her eggs.

The cardinal is a brightly-colored bird known for its whistling songs. It likes to build its nest near the ground in bushes. The nest, made of twigs and pieces of tree bark, typically holds two to four eggs that are bluish-white with brown specks.

Yellow warbler feeding its young

The female hummingbird builds a nest and cares for her eggs without the help of the male. First, she builds a cup-shaped nest made of moss and spider webs. Then, once she's laid her eggs, she keeps them warm. Hummingbirds usually lay only two tiny eggs at a time.

Most birds finish building their nest in about a week. If a nest is somehow destroyed and the female is ready to lay eggs, birds have been known to build a nest in as little as one day. Usually, a nest is used to raise only one brood. However, some birds, such as eagles, use the same nest over and over.

Not all birds build nests. Some, such as woodpeckers, peck out a hole in a tree. The woodpeckers then drop chips of wood into the hole as padding for eggs. Great horned owls have also been known to build nests in hollow parts of trees. Other birds, such as killdeers, dig a shallow scrape in the ground and lay their eggs there.

> *The season of spring is so named because it is the time of year when new life springs forth.*

Some birds may lay only one or two eggs, while others may lay more than 10. The number of eggs is usually related to the location of the nest. Nests in an area that is relatively safe will generally contain fewer eggs. A nest in an area with many enemies will have more eggs. Birds lay enough eggs to make sure that some of their chicks will reach adulthood; if an enemy eats some eggs or kills a chick, others will be left to grow up.

The color of a bird's eggs depends on the type of bird and the location of its nest. Nests that are well-hidden or inside a hole are

Young screech owls in nest hole

often filled with white eggs. A nest that is out in the open may hold dirt-colored eggs or eggs that are striped or spotted to help them blend in with the surroundings.

Spring comes to an end on the 21st day of June. Although many eggs will have hatched before then, birds that build their nests late in the season may not see their hatchlings until summer. By then, the days will be longer and warmer. Plants and trees will be green again, and the land will be full of worms and insects to feed hungry young birds.

North American birds that like to nest in mountains can choose from 11 different ranges.

Wood ducks hatching

INDEX

bluebirds 5, 13
blue jays 10
bobolinks 10–11
cardinals 18
chickadees 8
courting 7, 10–11, 12
    birdsongs 7, 10–11
    displays 7, 11, 12
daylight 4, 23
eagles 8, 12, 19
    bald 12
eggs 11, 12, 13, 18, 19, 20, 23
    color 20, 23
feeding 7, 10, 23
flycatchers 5
geese 5
hummingbirds 19
killdeers 6–7, 11, 20

migration 4
nests 4, 8, 10, 12, 13, 16–19, 20
    building 13, 16–19, 20
    location 13, 16, 20
orioles 10
owls 20–21
    great horned 20
    screech 20–21
predatory birds 8
robins 7, 10, 16–18
ruffed grouse 12
territories 4, 8, 10, 11
    size 8, 10
vireos 4
woodcocks 7
wood ducks 23
woodpeckers 12, 20
yellow warblers 18–19

J 598.156 MAS
Maslowski, Steve.

Birds in spring.

Smart Apple Media, c2002.

GLEN ROCK PUBLIC LIBRARY, NJ

3 9110 05054864 5

Glen Rock Public Library
Glen Rock NJ 07452
(201) 670-3970

NOV 2001